Hot Pink Flying Saucers and Other Clouds

FROM THE CLOUD APPRECIATION SOCIETY

Edited by Gavin Pretor-Pinney
Founder of The Cloud Appreciation Society

with Ian Loxley
Photo Gallery Editor of The Cloud Appreciation Society

www.cloudappreciationsociety.org

A PERIGEE BOOK

Introduction

The Cloud Appreciation Society was formed in the summer of 2004 in order to fight the banality of 'blue-sky thinking' and to remind people that clouds are one of Nature's most beautiful and life-giving phenomena. The society now has a membership in excess of 8,000, spread across 43 countries around the world.

Over the years, these members have sent us thousands of cloud photographs, which we've compiled into an online Cloud Gallery. We love receiving them and, before putting them up, classify them according to the particular cloud type and location: 'Cumulus' over Canberra, 'Altostratus undulatus' over San Francisco, 'Cumulonimbus incus capillatus' over Nairobi. The cloud formations are invariably beautiful, often dramatic, and, every now and then, extremely rare.

But the formations we enjoy receiving the most are ones that look like things: 'a cloud in the shape of a dog barking' or 'a flying-saucer cloud'. They are perhaps the rarest of the lot. Not only did someone have to be looking up just at the right moment, and happen to have a camera with them at the time, they also needed to be in the particular frame of mind required to be able to see shapes in the clouds.

Our extensive and rigorously scientific research has shown us that people tend to see shapes when they are cloud gazing in a relaxed and contented manner—when they have time on their hands and are not particularly bothered whether they find a shape or not. Only once they are delighting in the very formlessness of the clouds will they notice that, over by the trees, there's one in the shape of Salvador Dalí or a bear juggling a duck. This is the playful side of cloudspotting, which is why children are always so good at it.

For ages, we've thought that a collection of these photographs would make a great book. So we're very grateful to all the members who have allowed us to include their images, and very sorry to those whose didn't make this edition. With so many fantastic candidates to choose from, I really wish this book had room for more.

We hope *Hot Pink Flying Saucers* will encourage even more people, be they children or adults, to find a few moments in the day to gaze up, let their imaginations wander among the shifting shapes of the sky, and enter a world where monkeys go skiing, the mermaids are multicolored, and the Michelin Man decides to rob a bank.

Gavin Pretor-Pinney
Somerset, April 2007
www.cloudappreciationsociety.org

"O! it is pleasant, with a heart at ease
Just after sunset, or by moonlight skies,
To make the shifting clouds be what you please"

Samuel Taylor Coleridge
Fancy in Nubibus (or The Poet in the Clouds), 1819

A dog barking

or Cumulus congestus

Spotted over Soldiers Point, Australia
by **Terry Linsell**, Member 1531.

A dragonfly

or Cirrus and a contrail

Spotted over Upton, Lincolnshire, UK
by **Ian Loxley**, Member 1868.

The Michelin Man goes to rob a bank

or Cumulus

Spotted over Houston, Texas
by **Jason Tack**, Member 7413.

A red snapper

or Cumulus fractus

Spotted over Glencoe, Scotland
by **Andrew Edwards**, Member 8018.

The Cloudship Enterprise
or Altocumulus lenticularis

Spotted south of Calgary, Canada
by **Aidan McKeown**, Member 3139.

The Grim Reaper

or Stratocumulus with crepuscular rays

Spotted while flying over Kuala Lumpur, Malaysia
by **Richard Unwin**, Member 1945.

A first kiss is ruined by someone smoking

or Cumulus and Cirrocumulus

Spotted over Warrington, Cheshire, UK
by **Samantha Hall**, Member 3198.

Fusilli pasta

or Altocumulus undulatus

Spotted over Sossusvlei National Park, Namibia
by **Fabian Klenk**, Member 3485.

A golden eagle coming in to land

or Cirrostratus and Altocumulus

Spotted over Gaborone, Botswana
by **Arun Padake,** Member 5348.

A skateboarder doing tricks

or Cirrus

Spotted over Provence, France
by **Daniela Sarolo**, Member 5140.

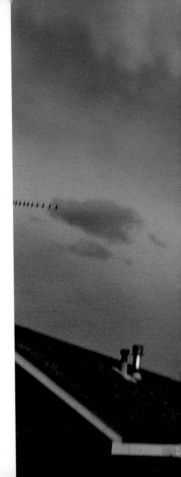

A storm cloud that has fallen asleep mid-shower

or Cumulonimbus

Spotted over Spixworth, Norfolk, UK
by **Steve England**, Member 6548.

Pegasus, the winged horse

or Cirrus lenticularis and Cirrostratus

Spotted over Antarctica, on a journey to the South Pole
by **Neville Shulman**, Member 2761.

A bear trained in the cruel sport of duck juggling

or Cumulus congestus and fractus with crepuscular rays

Spotted over Zagreb, Croatia
by **Sena Zutic**, Member 5882.

A flying saucer

or Altocumulus lenticularis

Spotted over the highway between Haleb and Damascus, Syria
by **Gianandrea Sandri** and **Roberto Cavallini**, Members 6681 and 6682.

A poodle coming home from the doggy salon

or Cumulus

Spotted over San Antonio, Texas
by **Sauni Coon**, Member 7142.

Halloween face

or Stratocumulus with crepuscular rays

Spotted over Nash Wood, Presteigne, Radnorshire, Wales
by **Howard Kirby**, Member 2163.

Salvador Dalí

or Cumulus congestus

Spotted flying over Nizza, France
by **Annegret Richter**, Member 7292.

A loggerhead turtle

or Cirrocumulus and Cirrostratus

Spotted over Aberdour, Fife, Scotland
by **Margaret Worrall**, Member 3381.

A fox terrier viewed from behind

or Stratocumulus

Spotted over Little Barrier Island, in the Hauraki Gulf, New Zealand
by **Louise Cotterall**, Member 3472.

Chief Rising Cloud

or Cumulus congestus

Spotted while flying over Lower Saxony, Germany
by **David Stolz**, Member 5635.

The Hunchback of Notre Dame declares his love for Esmeralda

or Stratocumulus with crepuscular rays

Spotted over Lista, Farsund, Norway
by **Ole Johnny Nilsen**, Member 2646.

A dragon's head

or Stratocumulus and Altostratus

Spotted over Clapham, Yorkshire, UK
by **Felicity Norman**, Member 8027.

A submarine

or Cumulus

Spotted over Teignmouth, Devon, UK
by **Kelly Rendall**, Member 6355.

Hidden faces of the sunset spirits

or Altocumulus

Spotted over Christchurch, New Zealand
by **Neil Wyatt**, Member 4518.

A monkey skiing down a blue run

or Cirrus

Spotted over Dunedin, New Zealand
by **John Lamb**, Member 1478.

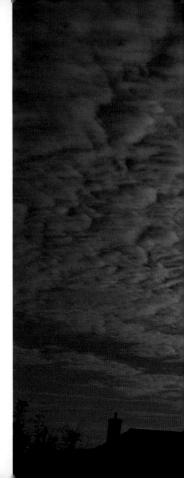

The number two

or Altocumulus with fallstreak holes

Spotted over Stafford, Staffordshire, UK

by **Richard Baker**, Member 1763.

A mermaid swimming under a giant frog

or Nacreous

Spotted over West Sands Beach, St. Andrews, Fife, Scotland
by **Andrew F. Casely**, Member 2516.

A proud mother greets her daughter
after the first day at school

or Stratocumulus perlucidus

Spotted over the Catskill Mountains, Monticello, New York
by **Jeffrey Pflaum**, Member 8019.

Index of Clouds

Index of Photographers

All photograph copyright remains with the photographers.
See page 64 for website details.

A PERIGEE BOOK
Published by the Penguin Group
Penguin Group (USA) Inc.
375 Hudson Street, New York, New York 10014, USA
Penguin Group (Canada), 90 Eglinton Avenue East, Suite
700, Toronto, Ontario M4P 2Y3, Canada (a division of
Pearson Penguin Canada Inc.) · Penguin Books Ltd., 80
Strand, London WC2R 0RL, England · Penguin Group
Ireland, 25 St. Stephen's Green, Dublin 2, Ireland (a division
of Penguin Books Ltd.) · Penguin Group (Australia), 250
Camberwell Road, Camberwell, Victoria 3124, Australia
(a division of Pearson Australia Group Pty. Ltd.) · Penguin
Books India Pvt. Ltd., 11 Community Centre, Panchsheel
Park, New Delhi—110 017, India · Penguin Group (NZ), 67
Apollo Drive, Rosedale, North Shore 0632, New Zealand
(a division of Pearson New Zealand Ltd.) · Penguin Books
(South Africa) (Pty.) Ltd., 24 Sturdee Avenue, Rosebank,
Johannesburg 2196, South Africa · Penguin Books Ltd.,
Registered Offices: 80 Strand, London WC2R 0RL, England

HOT PINK FLYING SAUCERS AND OTHER CLOUDS

Copyright © 2007 by Gavin Pretor-Pinney

PRINTING HISTORY
Originally published as *A Pig with Six Legs and Other
Clouds* in Great Britain by Hodder & Stoughton in 2007.
First American edition: November 2007

Perigee trade paperback ISBN: 978-0-399-53411-9

PRINTED IN MEXICO

10 9 8 7 6 5 4 3 2 1

Websites and other details of some of the contributing members:
ANDREW EDWARDS has www.lightandsky.co.uk; SAUNI COON would like to point out that her poodle
cloud was spotted by her son Ethan Coon; SAMANTHA HALL is Editor in Chief, *International Journal of
Meteorology* (www.ijmet.org); HOWARD KIRBY is Photo Editor of the *International Journal of Meteorology*
(www.ijmet.org); FABIAN KLENK has www.odd-fish.de; IAN LOXLEY is Tornado and Storm Research
Organisation Executive and Gallery Archivist, www.torro.org.uk; JEFFREY PFLAUM would like to thank
Michael Zwiebel for his scanning help; NEVILLE SHULMAN has www.shulman.co.uk; and
SENA ZUTIC has www.pticica.com/galerija.aspx?korisnikid=375.